Once a Man, Twice a Boy

MINES WERE: DEEP, DARK, DAMP, DUSTY AND DANGEROUS; BUT REGARDLESS OF THE HARD WORK, THERE WERE TIMES FOR FUN.

By
David Kuchta

Kiwi Publishing
5 Tippets Road
Nesquehoning, PA 18240

Photograph contributions by
George Harvan and Dan Farber

Second Printing
October 2002

Editor: Joy E. Kovalycsik
LC&N Research: Harry Murphy
Printer: Rush Printing, Tamaqua, PA

ISBN: 0-9649861-8-3
LC: 99-96344

No. 9 Wash Shanty Coal Museum, in Lansford, PA. This is one of three wash facilities built in 1924-25 for 50 thousand dollars. This was considered a very modern wash and change facility for the miners at that time. It now contains mine artifacts, tools, maps, books and a large display of models of the early L.C.& N. coal breakers

ONCE A MAN, TWICE A BOY

When most of the early immigrants came to America, they didn't understand nor speak English. One of the first jobs they received was in the coal breakers, picking slate, rock, wood or any other debris that came out of the mines. Of course, if they were miners in the old country, they usually were offered jobs in the mine.

In modern times, these coal breakers are now called coal preparation plants.

Many children of the immigrants also ended up working in these coal chutes. If a miner was killed, his son would usually be offered a job working as a "breaker boy." The first breakers built were, " Dry Breakers." They used no water until the late 1860's. These were tall wooden structures and were very dusty and noisy when they operated.

When the coal was brought out of the mines, it was transferred to the top of the breaker through some form of tipple. The mine cars had an end that opened up and when the coal car was hoisted to the top of the breaker, it was tipped in such a position that the load of coal dumped out the end of the car down into various chutes. Also in the breakers were crushers and screens. The crushers were to break the larger lumps of coal into smaller sizes and then the screens separated the various sizes of coal for market.

When all this large machinery worked the whole building shook and made all types of loud noises. About 1920, most of the old wooden breakers were either burned or tore down and replaced with steel and concrete buildings. These were still huge in comparison to today's coal preparation plants.

Most of the early "slate pickers, or breaker boys," were from 7 to 14 years of age. In the same case, they also had adults that did these same jobs. In the No. 9 breaker in Lansford, PA, it had about 50 slate pickers and half were adult and half were children. In 1914, the Federal Government passed the Child Labor Laws, and after that you had to be at least 14 years old to do any breaker or mine work. Of course, most of the later breakers were more automated and didn't need slate pickers anymore.

The early "Slate Pickers," sat on a wooden plank that faced what seemed to be an endless stream of coal passing in front of them or under the plank that was their seat.

They sat there for hours, bent over, removing the impurities that were mixed in with the coal. It was a monotonous, backbreaking job and was very painful on their fingers. Most of the younger kids couldn't afford to buy gloves and if they did, the sharp anthracite coal would shred the material in a very short time.

The breaker boss stood over them all day long. He usually carried some type of stick or cane, which he wouldn't hesitate to use on the young "slate pickers." He also would give them a good poke with the cane, if they left too much slate or rock pass by. If they dozed off, he would give them a jab to bring them back to reality. If he got any resistance from the young lads he

would shake it at them vigorously to settle them down. Of course, if they gave him too much trouble he just fired them on the spot.

Being this money helped support his father's income, none of the young workers wanted to lose this job. It certainly wasn't the best of jobs, but was a job never the less.

In the mid 1800s, the "breaker boys," got paid five cents an hour. They usually worked a 10-hour shift. In 1913, mine labors received 10 cents an hour. As they picked out the slate, rock or wood that came down the chutes, the impurities would be thrown into a special chute. These impurities then went down into a hopper or loading bin where they were then loaded into wagons, or in later times, mine cars and taken out to a rock or culm bank where it was disposed of. In the early days, the dust and noise was certainly detrimental to their health. Through the years some of the young workers, through carelessness, got caught in pulleys, gears or other mechanism.

In one state mine inspector's report it mentioned that the "coal breaker," stopped for a minor repair job and in the process the slate pickers started horse playing and one of the boys threw one of the other worker's cap down one of the chutes. While retrieving it, the machinery started up and the breaker boy was caught and crushed in some of the gears.

Some of the breaker boys were always looking for ways to get even with the breaker bosses and pulled off many pranks on them. But, it had to be done in a very clever fashion, because if he caught anyone doing them, they would be fired. No, if, ands or buts! All the boys looked forward to the day that they would be able to get jobs working inside the mines. The first jobs inside the mine were usually a door tender or mule driver. Later they moved up to mine labor and then eventually the most sought after job, that of miner.

If the miner became crippled, or too old to do his job, many times he returned back into the breaker and once again, became a "Breaker Boy." Hence: "Once a man, twice a boy."

Breaker Boys working in the coal chutes of a "Coal Breaker." (coal preparation plant)

Breaker Boys working in the coal chutes of a "Coal Breaker." (Coal preparation plant) This job was very hard on the hands of the workers. Gloves, as cheap as they were, were still too expensive for the young workers and would be worn out in a very short time from the sharp coal. During the mid 1800s, (Breaker Boys) only received five cents an hour. In 1912 they received 10 cents an hour.

INTRODUCTION:

As a writer and volunteer mine worker in the #9 Mine in Lansford, PA I became very interested in all aspects of mining. I researched and read all the various books on coal mining including all the State Mine Inspectors books on accidents and deaths.

Many of the mine books explain the different procedures and aspects of coal mining. As I interviewed various coal miners I found out that there is another side of coal mining that hasn't been printed or published or for that matter, never spoken about.

Coal mines are dark, damp, dusty, deep and dangerous. But as bad as these conditions are, the miners found time to fraternize with each other and also to have a little fun. Yes, they also pulled pranks on their buddies, as well as finding time for some horseplay. There was a time for work as well as time for play.

I also heard throughout the years, how various miners as well as company personnel cheated or took kick backs or pay offs. This was something that was never discussed in public. I tried to give the public a little understanding on how some of these acts were perpetrated. Every time I heard someone talk about the mine mules, I became very intrigued and wanted to find out all there was about these magnificent animals. I soon found out that there is very little written about the mules, and yet if it weren't for the mules during the early years, mining would never have progressed at the speed that it did.

After reading this book ... you will know the rest of the story!

Everyday Dangers To Miners:

(This was taken from an early State Mine Inspectors Report)

Mining of coal is a very dangerous vocation and the lives of the men engaged in it are always in jeopardy, and yet, the exercise of care and judgment on the part of both employer and employee would eliminate much of the danger. It is true that the managers of the mines of Pennsylvania have adopted rather stringent rules in their efforts to safeguard the miners, and have spent large sums of money to make the mines safe. But the fact still remains that the chief object of the management is to produce the greatest amount of coal at the smallest cost. That dear reader is the bottom line.

In the latter part of the 1800's, mine accidents of every description were common place. A big problem was the mere fact that many of mine workers were immigrants from various countries. Many didn't understand the English language. Some misunderstood certain orders or commands. Some early workers, didn't want to appear too ignorant, and just said yes to certain orders from various mine bosses, but in all really didn't understand what the mine boss was trying to imply.

The following are some excerpts of complaints from mine inspectors back in 1890.

I deem it my duty to make a few remarks on several of the fatal accidents that occurred in this district during the year 1888.

While it is a well known fact that all the accidents in and about the coal mines cannot be avoided, yet at least two thirds of the accidents in this district can betrayed to the neglect of the unfortunate victims themselves, or the negligence on the part of their co-workers. It seems that no law can be framed that will compel these people to take the proper care of themselves. As it can be seen from the facts stated; they almost seem to be bent on their own destruction, while in the majority of cases a little care and prudence on the part of the victims would probably have saved their lives.

The papers in the large cities and, in fact, some in the mining towns (the editors and writers being ignorant of the methods of mining anthracite coal), are constantly censuring the operators and their officials for their neglect, for not taking the proper care of the lives of their employees. They often pounce upon the mine inspectors for being too lenient in not properly enforcing the laws.

I have no hesitation in saying that I believe there is no country, or part of a country, in the world where the mine law is better enforced, and where the companies comply more willingly, than in the anthracite coal mines of Pennsylvania. The miners of anthracite coal are running more risks daily than any miners in this country, and perhaps, in the world, consequently the accidents are more numerous in proportion; yet they should be greatly reduced with the proper care on the part of the employees. If the papers editors or writers are really in earnest about the welfare of these poor unfortunates, who are each year mangled and crippled in our coal mines, they can do it much more effectively, by reasoning with the men. They can show them the great need of care and forethought on their part while engaged at this dangerous occupation. They should impress in their minds that they and they alone, are the only parties that can take proper care of them selves.

Here is a part of another mine inspector's report that was a little brutal toward the dead or injured miners.

Recklessness on the part of employees, or a lack of that rigid discipline which is just as indispensable in and about the coal mine, as it is for the safety of those engaged on railroads or upon the sea, should be reduced. It seems many victims contributed largely to their own destruction. There is no question of doubt; had one of the unfortunates contemplated taking his own life he could not have taken a better method. We have also stated elsewhere that according to the evidence of the foreman in charge of these collieries that several of the accidents are attributed to disobedience of orders given.

Here is a situation where this has happened. While working in an areas that were ready to blast, some workers will get themselves clear of the blast area while some will fail to go far enough and get struck by a flying fragments.

I know it to be a fact that it is a hard matter to get these Hungarians, (Slavs) into a place of safety while blasting. (From a report in the late 1800's). It seems that each of them thinks that they can dodge the flying missiles, but sometimes, they make a sad mistake and get themselves killed. However, it is a well-established fact that as a general rule, that insubordination is the result of lax discipline. It should also be understood, by those in charge of the mines, that as long as it exits, they must expect the same results.

Portal of the No. 9 Mine in Lansford, PA, in the year 1887. In 1931 the mine entrance and several hundred feet of the main gangway were rebuilt with concrete and steel collars.

No. 9 Coal breaker (coal preparation plant) built in 1870 shut down in 1908. Located north of the #9 Mine. In 1909, all coal from the No.9 mine was sent to the No.8 breaker in Coaldale, PA.

DISCOVERY OF COAL IN PENNSYLVANIA, HISTORY OF THE NO. 9 MINE

Anthracite Coal was first discovered here in Summit Hill in 1791. This fact is correct as far as Anthracite Coal being discovered in the southern hard coal fields, but Anthracite Coal was already discovered in parts of Lackawanna and Wyoming County in Pennsylvania. Some over enthusiastic historians have described this event as the first discovery of Anthracite Coal in Pennsylvania, but researchers claim otherwise. Even the very stories about Philip Ginder's life are erroneous. As early as 1826, a Quaker physician by the name of Dr. Thomas C. James read a paper about a brief account of the discovery of Anthracite Coal, before a Historical Society of Pennsylvania, in the city of Philadelphia. Even as he made his speech, he apologized for his problem of forgetfulness, because of aging. It seems from that time on the story that the good Doctor gave was the accepted version. Even Philip Ginder's name was misspelled. That is why some of the stories, statues and what have you offer the name Philip Ginter instead of Ginder.

Yes, Philip Ginder discovered Anthracite in 1791, but it was already discovered in Wyoming County, many, many years before. The earliest reports of anyone knowing about the existence of coal beds in Pennsylvania was by the Indians, 75 years before the Revolutionary war, according to William Griffith, Geologist and Mining Engineer from Scranton, Pa.

In 1710 Indians knew the existence of anthracite coal beds up in the Wyoming Valley area. On July 11, 1754, Lackawanna and Wyoming Valley coal regions were included in the sale of property by the Five Nations Indian Tribe to the Susquehanna Connecticut Co. The purchase price of the entire property was about $10,000. In 1762, the first Connecticut settlers, after camping at the mouth of Mill Creek, above Wilkes-Barre, and after several days of cutting hay on Jacob's Plains returned to Connecticut and reported the discovery of anthracite coal In 1763, Indians from that locality visiting the Governor and complained of being robbed of a mine by whites. This mine was believed to have been a coal mine. If you really give this a lot of thought, Indians were the first people who officially mined anthracite coal here in Pennsylvania. Without written reports or proof of this happening, the Indians once again will receive no credit, where credit is due.

In a letter from James Tilghman of Philadelphia to the Proprietaries, Thomas and Richard Penn, Spring Garden, London, it is stated that a Col. Francis had found a very great find of coal at Wyoming, in the hills that surround a very fine and extensive belief. "This coal," says the letter, "is thought to be very fine. With his compliments, Col. Francis sends you a piece of coal. This bed of coal situates as it is on this side of the river, and may at some future time or other may be a thing of great value.

In 1768 Charles Stewart surveyed Sunbury, and on maps indicated the presence of "stone coal" at Ross Hill, now Edwardsville.

Around 1750, Anthracite was supposedly discovered in East Central Pennsylvania, but it seems to center around a legend or folklore. It is a known fact that Indians were using stone coal many years before the White Man settled in Pennsylvania. They didn't use it for fuel but for other purposes. Another story was about some Indians seeking the services of a gunsmith to repair their flintlocks.

The gunsmith told the Indians that he had run out of charcoal and couldn't fire up his forge. The Indians supposedly left in a hurry, only to return a few hours later with a bag of what appeared as "black stones." By using a wood fire, the Indians ignited the "black stones," in the Black Smiths forge. The fire was so intense that the Black Smith was able to do his metal repairs. The Indians wouldn't relinquish any information as to the source of this mystifying fuel.

Like I said, this story is based on a legend. From what I have researched, I think this story is true. On August 14, 1766, a resident of Wyoming Valley mentions in a letter about an abundance of coarse, combustible material, which was, called Anthracite coal. The word Anthracite comes form the Greek work, "Anthrakitis." In 1769, two blacksmiths by the name of Obadiah and Daniel Gore of Wilkes-Barre obtained the location of a deposit of "black stones," from some Indians. In time they learned to use and maintain an on going fire, in their forges. In 1775, coal was being mined on the banks of the Susquehanna River, near Pittston.

By 1775, the news of this new heat generating substance spread to other Pennsylvania settlements and finally to the city of Philadelphia. In 1776, Two Durham boats located down the Susquehanna River, came up to the Wyoming Valley for coal. The coal was purchased from R. Geer, who had a mine opening about the mouth of Mill Creek. There are also reports that some Anthracite coal was used to help make arms for the Continental Army, in the governmental arsenal at Carlisle, Pennsylvania in 1775.

In the next 30 years, various experiments with the black stone were done but very little attention to it was achieved. At that time the use of coal was quite limited. In 1788, John Schopf, in a book named "Travels," mentions a visit he made in that year to a bed of brilliant black coal, a mile above Wyoming, which on handling leaves no taint and burns without emitting an offensive odor. Judge Jesse Fell, of Wilkes-Barre, Pa, made nails in 1788, with the use of anthracite. In 1790, coal was known to be plentiful in Schuylkill County, but was not mined because of a lack of demand.

In the fall of 1791, Philip Ginder discovered coal on Sharp Mountain near Summit Hill. Although little was known about this black rock that burns, Ginder had heard stories about it. The stories that said, Ginder was a poor trapper and hunter is false. Actually Ginder was a Miller and had built his own gristmill. This is the reason he was in the vicinity of Summit Hill. He was looking for sandstone conglomerate to make milling stones. In the process he came upon the so-called stone coal that he had heard about.

To satisfy his curiosity he took a couple of pieces to his friend, Joe Neyer that lived in the Mahoning Valley. Neyer was a blacksmith and was also inquisitive about this coal stone and threw several lumps on his charcoal fire to see what would happen. At first it didn't burn, but in time it ignited. Ginder's accidental discovery along with a successful experiment in Neyer's black smith hearth became well known throughout the area.

After Ginder realized you could burn the stone coal, he carried several lumps to Colonel Jacob Weiss. Weiss then took a couple saddlebags of this stone coal to Philadelphia were he showed the find to some prominent Philadelphians. These men then contacted experts who gave their opinions on the value of the coal as a mineral and its commercial potential. They found that they did have trouble keeping the coal burning, but after some experimenting they realized the value of this new fuel.

Colonel Jacob Weiss, along with Michael Hillegas, John Nicholson and Charles Cist, all of Philadelphia eventually purchased 10,000 acres of land were Ginder had first discovered the coal. They then formed the first coal mining enterprise in America, The Lehigh Coal Mining Co., in 1792. All this land that was purchased took in some of the area of where Lansford and the Number 9 mine is located. On February 11, 1808 Judge Jesse Fell made an experiment in the bar room of his hotel at the corner of Washington and Northampton Street and on the flyleaves of a book entitled, "The Free Mason's Monitor," he made the following memorandum:

"February 11th, of Masonry 5808, Made the experiments of burning the common stone-coal of the valley on a grate, in a common fire-place in my house, and find it will answer the purpose of fuel, making it a clearer and better fire, and less expense, than burning wood in the common way".

This experiment excited great interest and people came from far and near to watch further experiments, which were made in a grate of iron, and soon there were dozens of similar grates being constructed and put into use. From this point on, the use of anthracite coal took on a whole new prospective as another type of fuel to be used in forges, industry, and in homes.

****

In 1817, an engineer by the name of Isaac Able Chapman was hired by the state to do land mapping of certain regions. Along with his regular work he also marked down various outcrops of coal veins in, which later became, Carbon County. This is the same Chapman who drove the Hackelbernie Tunnel and may have been involved with the down track of the first gravity railroad which later was known as the Switchback Railroad that ran from Mauch Chunk to Summit Hill. Isaac Able Chapman also became the first engineer hired by the Lehigh Coal and Navigation Company.

First Mining In The Southern Coal Fields:

In 1814, Miners, Cist & Robinson made their first shipment of coal to Philadelphia from an area near Mauch Chunk. Two boatloads arrived at Philadelphia and were largely purchase by White & Hazard for $21 per ton and used in the first successful effort to utilize anthracite in the iron industry at their wire mills at the fall of the Schuylkill. In 1815, Being firmly convinced of the value of the coal, White, Hazard and Hauto set about securing a lease on the Lehigh Coal Mine Company land near Mauch Chunk, for 10,000 acres for 20 years, for "one ear of corn per year, if demanded."

In 1819 a coal quarry, which got to be known as "The Old Mine," in the Summit Hill area was started. This Quarry was located in the area of the "Mammoth Vein." This large Mammoth bed of coal got to be known as the most important of all Anthracite deposits. In 1822 the Coal Company became incorporated under the name of Lehigh Coal & Navigation Co.

As early as 1814, two prominent businessmen or entrepreneurs, by the names of Josiah White and Erskine Hazard were starting to get involved with Anthracite Coal. These men were very interested in promoting and the marketing of this new fuel. Because of them the building of the Lehigh Canal system came about. Also they were instrumental in building the first gravity railroad and the development of the vast coal lands between Tamaqua and Mauch Chunk. Their accomplishments could fill volumes of reading material.

During 1820, The Lehigh Coal and Navigation Company began mining and shipping coal. The price for a ton of coal at Philadelphia was $8.40. Shipment that year was 365 tons. 1829 opened The Lehigh Canal from Mauch Chunk clear down to Easton. In 1837, Lehigh River navigation to White Haven was opened. In 1831 the Nesquehoning Railroad and plane was built. By 1849, the quarry in Summit Hill had to be abandoned because of flooding waters. At that period of time, they didn't have the high volume water pumps and couldn't remove the water to keep the quarry operable.

In view of some coal workings being soon worked out and also the large quarry being flooded, it was decided to drive a tunnel on a level with the bottom of the Panther Creek Valley into the large (Mammoth Vein) on the south side of the valley. At this point, we can use the dates of 1844-45 as the time of the No. 9 Mines conception.

Around 1845, plans for the No. 9 Mine were on the drawing board. Two attempts to drive the mine, back in 1851 and 1853, were unsuccessful. In 1851 the Company set its goal to drive in around 800 yards to the Mammoth Vein of coal. On January 14, 1852, a proposal was accepted from Daniel Bertsch to drive the No.9 tunnel, a distance of, "about seven hundred yards." On March 26 of that year, the contract with Daniel Bertsch for driving the tunnel No.9 was ordered to be executed under Corporate Seal. In that year the tunnel was driven forward 166 lineal yards at a cost of $6,228.74.

In the Annual Company Report, in 1853 "But little progress has been made in driving forward the No.9 Tunnel, since my last Annual Report. It seems that after driving in about 195 yards, the rock suddenly dipped down below the bottom of the tunnel, leaving a bed of quicksand to contend with. After several attempts to secure it with timber, it was given up as impracticable, without incurring a large and unjustifiable expense. As a final resort, it was concluded to make an open cut, about 150 feet in length and 70 feet in depth, which will require three or four months to complete.

There has been expended on this project, $9,807.81."

By January 1, 1856, Tunnel No.9 had been driven forward 1,119 feet without striking any coal. This project took three or four months to complete this work. In time they also found that near the beginning of this mine was an area know as a "squeeze area," During the early years this had to be heavily timbered. It also should be noted that during the years prior to 1867, most mines didn't work during the winter season. This "down time" lasted approximately four months. In the spring, when the canal systems reopened for business then they started mining coal once again. . During the winter season, repairs and other work besides mining were done in the tunnels. This work was called "Dead work." The Company used a cut off date of Dec. 10th when all the Canal systems were shut down, and wherever possible all the canals were drained of water because of freezing conditions which would damage much of the canal structures. Because of the winter freezes and shut downs of the canals there wasn't any other way, up until that time, to haul the coal in larger quantities to the big city markets. Once Railroads were established, the mines started working year around.

During 1855 and 56, driving in the #9 mine had begun in earnest. The miners had reached the large Mammoth vein of coal, on December 25, 1857. The miners had to drive 2,283 feet to accomplish their goals. On January 30, 1858, a proposal for mining coal from the No. 9 Tunnel came from Daniel Bertsch for the years 1858 through 1860 for 90,000 tons of coal per year was accepted by the Company. This led to the first mining contract for this particular tunnel On March 31, 1858, the mining contract with Daniel Bertsch was ordered to be executed under Corporate Seal. In later years, the LC&N Company named a street after Bertsch in the town of Lansford, Pa.

CHANGE OF OWNERSHIP

The No.9 Mine operated throughout the years until 1954-55. The No. 9 Mine, along with all other deep mines that were operating throughout the Panther Valley Area, by the Lehigh Coal & Navigational Company shut down all operations.

In 1955, a group of individuals gathered together and formed, "The Coaldale Mining Co." In 1960,The Fauzio Brothers, bought the entire coal operations, from the Lehigh Coal & Navigation Company and called itself; "The Greenwood Stripping Corporation."

In 1962, a small group of miners formed themselves a small mining company called, "Lanscoal," and leased the Number 9 Mine from the Fauzio Brothers. This small operation ran the Number 9 Mine until June, 1972. At this point the mine was the oldest and longest operating deep coal mine in the United States. In 1974, the Fauzio brothers sold the property of their Greenwood Stripping Corporation, to the "Bethlehem Mines."

In 1989, Attorney James Curran from Pottsville bought the mining property from the Bethlehem Mines and went under the old name, "The Lehigh Coal & Navigation Company."

In 1990, "The Panther Creek Valley Foundation," (No.9 Wash Shanty Mine and Museum,) was formed with the idea of creating an organization whose main object was to preserving our local history. Their first objective was to create a library but as time went on the direction changed. They obtained a lease from Mr. Curran, for the old Wash Shanty building and surrounding grounds, to utilize the building as a museum for mining artifacts, equipment, pictures, historical maps and anything pertaining to mining. In August of 1994, Attorney James Curran, president of the Lehigh Coal and Navigation company gave permission to the Number 9 officers and volunteers, to realize their dream of the re-opening of the sealed Number 9 mine tunnel in Lansford, Pennsylvania.

Through the next six years, the mine volunteers while working every Saturday, and some Sundays, cleaned tons of mud and debris from the tunnel. Also, some sections of gangways that passed through veins of coal had to be re-timbered. The largest feat was driving a second opening from the main water level gangway to the surface. This project took close to 18 months.

The goal is to have the tunnel ready for tourism by the year two thousand. The No.9 Mine will have the original Dr. Young's Memorial Hospital Room which is carved out of solid rock. Also on display will be a working mine chute, the mule way, where they brought up the mules from the second level to the first level, and a mine shaft that will have the two original cages (elevators) locked into place on the water level. The original go-devils which are the equipment to load cars on to the cages will also be on display.

How the Lanscoal No. 9 Mine Came About. (The following article is reprinted from the Friday, February 15, 1963 issue of the Landsford Evening Record. It deals with the forming of the Lanscoal Number Nine Mine) By: Joe Matula

On February 29, 1960, the Coaldale Mining Co. terminated its lease with the Lehigh Navigation Coal Co., at the No. 9 mine situated at the West End of Lansford. This was the last underground mining operation performed there
until January 30, 1961, when four local men, idled by the shutdown of mining operations in the valley formed a cooperative concern incorporated as the Lanscoal Mining Co.

They received approval of the United Mineworker's of America to run a "Union Mine" and through the cooperation of James Fauzio of Fauzio Brothers, who own the Greenwood Mining Co., started production. The four officers of the firm and those responsible for the underground operation as it is today are John Zulu, president; Steve Carbines, vice president and general manager; Joseph Oriole, treasurer all of Lansford and Steve Zilch of Coaldale, secretary.

The operation started with 14 men and now employs 18. There are approximately 200 tons of coal mined a day in the operation at the "Water level." All told, with truck drivers and other helpers, there are 22 jobs involved in the operation that has provided a total of 239 days of work in 1961 and 284 days in 1962.

The four men sub-leased the property from the Fauzios, the prime lessee, and take the coal to the Greenwood firm for processing. The coal processed there is used for the local markets as well as shipment to different part of the country and overseas. A miner's workday starts at 7 A.M. and ends at 2 P.M. Electric mine locomotives are used by the miners to transport the coal out of the mine five days a week. The operation is unionized, with all benefits for the miners including death, workman's compensation, occupation disease (Anthrasilicosis) and vacation pay. There are state inspections made at the mine every two months, while federal inspection is made twice a year and insurance inspection is made once a year.

Six months of preparation were necessary by the people instrumental in the opening of the mine, prior to production. Many obstacles have been met in the process of operating the mine, but they have all been resolved. Some thought has been given to the opening of a second level of the mine, but the cost at this time is still prohibitive, although planning for the future includes this move.

As one walks into the mine, he notices recesses in the side of the walls and learns that these are know as "shelters." The miners use them when the locomotive is hauling the cars in and out of the mine. Going on further into the black depths of the mine, one sees a sign over a doorway that is familiar to the old miners. It states, "Dr. Young Memorial Hospital." This room located inside the mine, had many facilities for the treatment of mine accident victims and previously treated miners before they were taken by trolley car to the Ashland Hospital. This was used in the days before the Coaldale Hospital was erected. Even though it was located within the mine itself, it was kept clean and the walls were painted white.

Further inside the mine can be seen a red sign on a fence which designated an alternate rout out of the mine. This is one of several required by federal authorities since the Knox mine disaster. About 1,800 ft. inside the mine can be seen a picture of the Sacred Heart of Jesus, which was enthroned by the nuns of St. Michael's School, Lansford on February 22, 1962. Although the picture is curled due to dampness and water, most of the miners still stop to say a daily prayer prior to starting work in the morning. Many old shafts in the mine have been sealed off and new ones opened Some of the old tunnels are used to store empty coal cars and full one prior to removal by the locomotive. Timbers used in supporting the mine consist of two legs, which are the uprights and a collar, which is notched in such a way that the pressure of the earth above forces it against the legs. Each of the timbers weighs about 150 pounds and must be handled by the men themselves.

A "miners shack," located within the mine, contains a small heater on which several miners place potatoes in the morning. By lunchtime, they are well roasted. Chutes and walkways are located throughout the mine. Chutes are used to release the coal from the veins into the empty coal car. The walkways permit the men to walk the entire length of the chute to dislodge any coal that may bind there.

Often it is necessary to blast the rock and coal loose with dynamite. The blasting caps used to detonate the dynamite have a safety device on them as well as a break in the line used for the actual blast. Although many people have always been under the impression that the most dangerous job in the world is underground mining, safety precautions have been set by the state, federal governments, insurance agencies and the men as well to safeguard lives and minimize risks. It should be stressed that the four miners who put their money into the opening of the Lanscoal No. 9 mine placed themselves in a very dangerous financial position. They invested all of their money into the prospects of operating the mine for at least ten more years. They, as well as their crew, worked without any money coming in until they had re-timbered much of the mine, cleared the water runway, and generally had the facility to a point where it met safety regulations. This preparatory period, before a dollar profit could be realized, covered some six months. The men above worked for six months without pay, and when they started getting money from the coal they mined, they certainly didn't get rich. Most of these men were to young to retire, too old to get other jobs elsewhere. They had no choice, except to do the thing they knew best, and that was mining.

.

****

First, Lehigh Coal and Navigational Coal Companies open pit quarry (Mammoth Vein) located in Summit Hill, PA. in 1821 (Richard Richardson, Memoir of Josiah White, 1873

Breaker boys (and men) working in the coal chutes to remove any slate, rock or wood that is mixed in with the coal that comes out of the mines. In the old No.9 Coal Breaker, located north of the No. 9 Mine, they had approximately 50 workers in these-coal chutes. Half of these workers were men.

Sister M. Methodia, M.S.C.

Sister Methodia established a mini chapel in the #9 mine in Lansford Pennsylvania. It was dedicated to her patron; The Sacred Heart of Jesus. She cared so much for the coal miners that they made her an Honorary Member of the Last of the Panther Valley Coal Miners Club. The sister also had the name, "Sister of the Anthracite Miners."

It is said that once the shrine was consummated and blessed, no one had any bad accidents or injuries until the mine closed. This alone is a remarkable feat. At this shrine, the miners began and ended each day, pausing for a brief prayer as the entered and left the mine tunnel. This was the start to what became a series of regular visits, usually with several other Sisters, all wearing hard hat with a battery lamp, boots and coveralls over their habits, to pray with the men and present them with religious medals. Once every year, the miners would bring Sister Methodia and some of the other Sisters, from St. Michael's in Lansford, and hold a mass for the miners. They would sing hymns and say prayers. The miners looked forward to this each year. During the early 70's a Nun from the same order of Sister Methodia, was taking a tour of the No. 9 Mine. She wanted to see and bless a chute of one of the miners. As she went up into the chute, she had on her white habit, under a pair of coveralls. At one point the miner who was helping her along, told her not to step on the sheet iron chute, but to step over it. Well, she forgot herself and stepped on the slippery sheet iron and down she went. She fell about 30 or 40 feet.

Thankfully her guardian Angel was watching over her and she didn't get hurt except for some wear and tear on her habit. Luckily, she was wearing an outer covering of coveralls When Lanscoal had its final day of production, Sister Methodia was there to give her final blessings to the mine and the last employees who were working there. Her name and date is written on the wall where the Shrine was erected. Sister Methodia passed away at the age of 102.

The Final Days Of Lanscoal

Approximately five weeks before the Lanscoal Mine closed, Richard Benyo of the Lehighton Times News paid a visit to the Lanscoal mine. He was so inspired by what he saw that he wrote a lengthy article named, "Death of the Giant." May 13, 1972. I in return, have taken Benyo's article and have rewritten what I think he saw and felt, on that memorable occasion. I tried to see through Benyo's writing what had transpired in those last few weeks before the mine shut down.

The Last Farewell To The Matriarch Of The Panther Valley Mines (1972).

Under the hills and valleys of the Panther Valley, there are countless miles and a miles of tunnels reaching out in every direction. They are dark, damp and certainly massive. To some it might appear frightening, to others intriguing The Lanscoal mine is certainly magnificent. This mine is not only a giant among mines; it is the "Matriarch," of all Panther Valley Mines. At this time she is the oldest operating anthracite coal mine in this country. At times she is benevolent, sometimes cold. One fact remains...and that she is dying. She knows her time has come. With the impending death of this giant, an era will come to an end. Deep coal mining in the Panther Valley will be no more. Within a short period of time, the final word will be given, to shut down this old mine.

In the mines, it is not preferable to work the last day. You just don't know what could be in store for you. Mother Nature is one fickle lady. She has taken human sacrifices, down through the years, of those who dared take her riches that she carries in her bosom. When the miners met on those last days, they knew that the end was eminent. Some of these men worked years together. Some claim that they hate working in the mines, after all it is hard work, dirty work and at times there is a heavy toll to pay for those who would descend deep into the earth itself, to rob Mother Nature of her riches. They all know that in time they themselves will be robbed of their ability to breathe. But in the same case, they hate to see it all come to an end.

The men feel these strange emotions coming over them. It is strange and freighting. These are rugged men that have worked hard, and lived hard. They knew their jobs well. There is no complaining they just do their job, and do it right. There is no room for error. At the same time, there is some laughter, some joking going on. These men are taking one day at a time. The mines were scheduled to close over a year ago. They knew they were on borrowed time. They also know that their type of work will be no more. They know that deep mining will go the way of the dinosaur. They are but a mere handful of men, not quite ten, who are now the remains of a mining force of over 600 workers when this mine had the distinction of being called the #9 mine.

The mine only produces a mere 30 carloads of coal each day. In the past, car after car of coal came out of the deep recesses of this great "Old Matriarch," 24 hours a day. This day will be different...Richard Benyo was there to see this happen, and to write the obituary of this Grand Old Lady. He entered the mine to see what it was all about, before all deep coal mining becomes extinct in the Panther Valley. He will never forget that experience. It was both pleasant as well as sad.

As he entered the mine, he could feel the drip of the water falling on him. He could smell the musty odor that is so unique in it self. A smell so different than any smell that he has ever encountered. It's a mixture of rotting timbers, wet coal and rock, and the presence of blasting powder smoke. It is a strange smell, but yet not totally uncomfortable. In all reality, it is the perfume of which this grand old lady of the mines gives off. Perhaps today, it is a little different, she knows she is dying. The tunnel is foreboding, it is a giant that is on the verge of death. As he met with the miners, he could feel that certain tension that there is something different on this particular day. The giant knows that this man is different. She knows he doesn't belong. Benyo watches as the miners prepare the cut. It's totally different than what he has ever encountered before. The drilling of the holes, packing charges, setting off the charge and then the rush of coal down the chute.

This giant has given up its precious gems, once again, like it has done for countless years. This time it is four cars of "Black Diamonds," ready to be taken to the breaker for preparation. In about five weeks from this occasion, the giant will go into its perpetual sleep and be no more. Her exits will be sealed for posterity. Nature will once again take over what is truly hers. The men who tormented her will be gone. When the portal is sealed, she will no longer be the Matriarch of all the mines; she will be only a memory.

This giant has fed, clothed, and educated people from the Panther Valley for over 140 years. She has also killed and maimed her share, but she will kill or injure no more. Her time has come, to say goodbye.

Documentation of the closing of Lanscoal #9 mine, by George Harvan, Photographer:

From 1947 to 1955, George Harvan was a free lance photographer for the Lehigh Coal and Navigational Company. Since 1976, Harvan was chief photographer of the Valley Gazette, in Lansford, Pennsylvania. He is dedicated to the preservation of mining folklore in the Panther Valley. George Harvan is blessed with being a top rated photographer and having had the opportunity and ability to be able to document on film, the hard work of the Panther Valley miner. The miners placed their trust and friendship in him to such a point that they welcomed him to record their very lives while working in the mines.

Many of his documented pictures of certain coal activities are one of a kind. They would have never been seen, by the general public, if it wasn't for Harvan's willingness to go deep into the depths of the coal mine, up chutes were all the work took place and take these pictures. Miners were a special breed of worker. There never were conditions in the mines that could be considered good. Most of the time conditions were dirty, damp and dangerous. On occasions the dust was so thick, they you could cut it with a knife, but Harvan was there, recording the mining activity as it happened...all for posterity.

George Harvan has spent countless hours, scouring the mines, looking for the special pictures that have documented the various works done by miners for posterity. The miner's story was George's story. The only difference was that Harvan was in pictures. Thanks to him, people now know what deep mining was all about. Harvan captured that inner soul of the miners. Most miners paid a huge price for working in the mines. They put their lives on the line, every day when they entered the mines. Even those that were fortunate enough to avoid the tragedies caused by cave-ins, explosions or lethal gases knew the possibility that they would probably get the deadly Anthrasilicosis or "Black Lungs," which robbed the miner of his breath and at times his very life.

Lanscoal Mine in Lansford, Pa. (formally known as the LC&N Coal Company) #9 mine, had been documented as the oldest continuously operated Anthracite deep coal mine in the world.

June 22, 1972, became a very sad but historic date in the annals of Panther Valley Anthracite mining. This day was a very sad day for Harvan and the handful of veteran miners. After 150 years, the last coal cars of Anthracite were destined to leave the mine.

The last day's operation was supposed to be brief and simple. A string of coal cars were to be loaded with the "Black Diamonds," just as it had been happening for the last 150 years.

Mother Nature on that day had other plans. She also had the final say. By mid morning, Hurricane Agnes struck with a vengeance. Being #9 mine was the final Matriarch of all the Panther Valleys deep mines, she didn't intend to bow out gracefully.

Heavy rains started reaching dangerous depths in the gangways. As a result the tracks leading from the tunnel were obliterated with water and mud. In a short time, rails spread or broke free from their moorings, causing the loaded coal cars to derail, time after time.

The last major challenge of the Lanscoal miners was to get the final "Trip" of Anthracite Coal to the outside of the mine. Man and nature collided. Neither wanted to give in.

As the day wore on, the workers, although soaked to the skin, continued to work with water up to their knees. Finally, the last trip of coal cars, which was reduced to two cars, reached their final destination. A large sign on the rear mine car read: "Last car of coal mined in the Panther Valley."

This was the end of the Grand Old Lady of the deep coal mining industry in the Panther Valley area. Regardless of all the rain and mud, and under the worst of conditions, George Harvan was on hand to photograph it.

Time to say their good-byes.

When the last of the miners emerged from the mine, they huddled in a near by shed and at that point of times, it was hard to determine if their faces were wet with rain or with tears of sorrow. As the rains poured, certain sadness overcame the group as the men realized that this was the last day the men would work as a team.

At this point, Mike Sabron, a miner from Summit Hill spoke out: "We shouldn't forget each other. Lets have a get together once a year for old times sake." These very words gave inspiration of the remaining miners to form, "The Last of the Panther Valley Deep Coal Miners Club." For 26 years the organization of Lanscoal miners has reunited at a festive annual dinner meeting in Lansford each August. Their number in 1999 is reduced to four members.

Rats, The Miners Friend: Sort of?

In the Valley, rats were utilized as an early warning system, warning the miner that something is a miss. One story told to me was when a mine personnel was feeding three rats outside of a fire-boss's office, (inside the mine) The rats got a little excited and started running away. The miner also took off. A big rock fell in the area that the rats were being fed. The unwritten law of the mines was to never kill any rats. The reason for this is that rats also made borrows or tunnels in the floor or along the walls of the mine tunnels. If there were any squeeze of rock formations or shifting of the ground above, the tunnels of the rats would start closing in. They also claim that the rats had very sensitive ears. The rats soon found out that this meant trouble and would take off for safer places.

When miners saw the rats move out, they also knew it was time to get out of the area. Most miners learned to look at the top cut of the posts or legs that hold the collars. If they saw the pitch cuts closing tightly, they knew that the rock above is starting to shift. This was also an early warning system of oncoming danger. Miners ate their lunches in the mines. Usually there weren't any facilities to wash their hands. So when eating a sandwich, many miners used to throw the piece of sandwich that was blackened from their hands, to the rats. The rats depended on these handouts for food. Rats also cleaned up human excrement that was deposited in the mine ditches. Yes, when nature calls, the open ditches along the rails were used. Usually there was a good flow of water and being the water was more acidic, it broke down the human waste. Now in modern times, the laws say that regular toilet fascinates must be in place inside the mines. Believe it or not, they are built over the ditches, where all the wastes end up

In 1891 there was a very bad mine flooding up near Hazelton. After 20 days five men were found alive in a breast. Eventually six bodies were also found, but rats ate them almost beyond recognition. A rat that is cornered or threatened can be very aggressive. So most miners had a philosophy of leaving them alone, and they in return, will leave you alone. Some miners didn't always appreciate all the rats. One fire boss had a cat or two, to keep the rats out of his office. It seemed that some of the rats took to his office and would shred his papers to make their nests.

Also some mule tenders were known to set up traps that would drown the rats, because the rats would get into the mules oats and cause problems. Hungry rats at times would invade the underground mule stables and eventually chew holes into the mules feed box. Sometimes the mule would try to push the rats away with their nose and in return some rats would bite the mules around the nose or mouth. Overall, the rats were protected and respected. Doing volunteer work in the No 9 mine for over a year, there are no rats because of a lack of food for them. When we volunteers break bread, we leave the mine and head to the Wash Shanty Museum, where one of the museum volunteers of the foundation has some type of hot meal from stews to pizza. Whatever it is, it's always good.

Bodies Coming Down The Coal Chute:

This incident happened in a mine in Lansford, PA. At the beginning of the shift, one of the miners thought they would have a little fun with another worker called. Hokey Pokey. This man worked in a circus or carnival and was sort of a magician. This is the way he got his name. Just before they were going to load a mine car that was placed at the bottom of one of the chutes, one of the miners told Hokey that he thinks the coal chute was located under a cemetery and that if they weren't careful they might pull in the graves above them.

Unknowing to Hokey Pokey, the miner placed a dummy above the chute board that was located in the battery above the mine car. When Hokey removed the chute board, down came the dummy. When Hokey Pokey saw this he took off down the tunnel, and then took off for home. Everyone had a good laugh except the supervisors. The man that perpetuated this prank was almost fired. It goes to show you that miners did have a sense of humor, although at times it did go too far.

Pranks And Horseplay:

It's something that most miners don't want to talk about. But when you got some talking they finally spilled the beans about the different pranks and horseplay that they saw or did. It did seem that nailing someone's lunch can down to a wooden floor was a universal joke. It also happened around the mines. Before quitting time, someone would nail the can to a wooden floor. The owner in a hurry to get on home, would grab for the handle, only to find the handle breaking loose form the can, and the can staying put. Maddening; You bet! Another on going joke was to smear some Limburger cheese on the resisters of the mine motor car. After a while it would smell to high heaven. Some guys also like to smear Limburger cheese on the inside of the miners helmet when he had the helmet setting aside. A real dirty deed was when someone was taking a nice hot leisurely shower, and someone would go outside in the winter time and get a shovel full of nice new fallen snow, and then throw it on the miner.

Stanley V. Stanek remembers when some of the guys liked to put some paper in the toe of the miner's boot. Not quite enough that they can't get their foot in, but enough to pinch the hell out of their toes. The miner upon finding this, didn't blow his cool, all he said is, "I guess boys will be boys?"

Another universal joke was in the showers, when someone was soaping up, and have their back turned, they would turn on the cold water. It would take the breath away.

Joe Kraycirik of Lansford told me that he was always afraid of the mules. I sort of get the idea that he just didn't trust them. One of the mule drivers by the name of Steve Lewis knew of Joe's fear or mistrust of the mules. If Lewis saw Kraycirik ahead of him he would hit the mule in the backside with a sprag. Then the mule would run toward Kraycirik, following his mining light. At this point Joe would head up the nearest mine chute he could find, only to see Lewis laughing his head off. Kraycirik also told of another story that seemed to be passed down in time. It seems that an old Polish miner walking in the main tunnel, with a long metal bar or hand drill, accidentally hit the hot line and got knocked on his rear.

When he got up, he was so mad that he took another swing at the hot line only to be knocked down the second time. At this point, he learned his lesson. Many miners would buy a set of work clothes and never take them home to be washed. Some would wear them until they fell apart. When you had these clothes hanging from the basket next to yours, the smell was very bad. In a book by George Harvan, "The Coal Miners of Panther Valley," Harvan explains it all. He says that the wash shanty trademark was a strong, pungent odor produced by clothing worn on the job. I can relate to this because I too, changed work clothes for 40 years, and had my clothes hanging in the same manner. Harvan goes on to explain the various smells from, acid water, decaying timber, dynamite smoke and sulfur fumes which penetrated the clothing fibers and resulted in the stench of the wash shanty. The one smell he didn't mention is the smell of perspiration and bad foot odor. I think most miners could live with the smell of the mines, but when it came down to personal hygiene, then it was a different story.

Some guys would get a long pole after the miner was gone, reach up and bring the pants or shirt on down and then tie multiple tight knots into the shirt sleeves or the pants legs. The next morning the miner would be fit to be tied. Sometimes he got the message, some times he just couldn't care less. This particular story wasn't a prank but when it happened everybody was concerned, but when they found out the guy was all right, then they all had to laugh. Of course the "Patcher," didn't think it was so funny because it happened in wintertime. It seems that a sump was being dug in the ditch along side the rib. Upon going into work at 3:30 in the morning, the patcher walked along side the electric motor, in the ditch. He stepped into the sump hole and went clear up to his chest in water. Not suspecting this, it scarred the heck out of this guy He wasn't all too happy about the incident and madder because everyone laughed about it!

One day, the guys who worked down in Lanscoal mine thought they would have a little fun with big Mike "Tarzan" Lukas. Someone took a deer head all the way up into the face of his chute. When "Tarzan went up into the chute, the light from his miners light hit the glass eyes of the deer which shown back. "Tarzan came running back down the chute hollering about something up there. Tarzan was hollering that a couple of owls were after him. The mine crew all had a heck of a good laugh over that one.

When the mines were all working, the miners down in the 3rd or 4th level, thought they would have some fun with Andy Maholick. One of the guys took up an mannequin, all the way up in his chute. When he saw it, he came running down. He couldn't believe his eyes. I guess he thought it was the real thing. This happened over 35 years ago. Mike "Pussy" Paslawsky was a big miner. He feared nothing, but Mike Sabron knew his weak point. One day he looked at "Pussy," and told him he don't look so good. That he looks kind of pale. Before you know it, "Pussy" went to Zuzu and told him that he wasn't feeling so good and was heading on home. I guess we can call this the power of suggestion.

Stealing Chewing Tobacco:

One of the mineworkers was known to be stealing different guys chewing tobacco. They wanted to teach him a good lesson. So after one of the mine vacations, they rounded up some Mule crap that was all dried up. They mixed it in with some Elephant Butts Chewing Tobacco and put it in one of the miners coats.

When this guy went into the coat pocket and took out the Chew, he took a large mouthful. It wasn't too much longer that the guy got a little green under the collar and was complaining that he didn't feel too good. I think he learned his lesson. Some miners were also known to go into other guys lunch buckets and help themselves to a sandwich or two. At that time, the inner tubes of trucks were a red color. They would take a piece of the red inner tube and put it between the salami or pressed ham. When they chewed on the rubber inner tube, they knew something was amiss. Some even put new condoms in the sandwiches. "Dutch" Yates from Summit Hill remembers when one of the miners stayed out half of the night boozing. When he got home, his wife was quite upset. She knew hollering at him wouldn't do any good, so she fixed his wagon another way. When the miner stopped for his lunch he had one hell of a surprise. His wife put beer bottle caps in his sandwich. She got her revenge.

I had to give the following story some thought, but I decided on telling the story. Believe me, it's a true story.

Most miners only had the lid of their lunch cans filled with water. Some took a small container of water to work. It was just enough to wash down the lunch and quench the thirst. Some miners that drank too many brews the night before, and had one hell of a dry-on the following morning. No matter how much water they drank, it was never enough. So some guys took to bumming water from their buddies.

Some asked; some stole. One fellow was known to help himself to everyone's water. So the one day, one of the guys thought that he would fix this guys boots. When they came out of the mine, the fellow made sure that this water thief was watching. He relieved himself right in his water container. He then went to a outside water hose, rinsed out the water container and then filled it with water. He approached the water thief and asked him if he wanted a drink.

This guy thought he flipped his lid, but the one thing this water thief never did again...was to drink from this guys water can.

Kicked by a Dead Mule:

One of the mine mules had broken its hip. The mine foreman called out and had one of the Companies Veterinarians (Dr. Weaver) come and give the mule a shot, to put it out of its misery. Kraycirik remembers staying overtime to load up the mule on a flat mine car to take it out of the mine. As they went to put it on a truck, the hoof got stuck on a nail. Joe was up front by the head and his buddy was at the other end. When his buddy grabbed the tail to move it, the hoof let loose and caught Kraycirik smack in the belly. It scared the hell out of him!

Billie Winebelly:

Another story that has been passed down the line was a fellow who had the job that started earlier than the rest of the miners. He got into the habit of coming about four hours early and then would lay down on a pile of planks to take a little snooze. In the morning, the "Patrolman," whose job it was to check the mine out for any gas or other unsafe problems. When he passed, "Billy Winebelly," as he was known, sleeping there, he hollered out to him, "Hi there Billy Winebelly," which awakened and scared the hell out of him. I guess he thought the devil was after him.

John Lennon remembers working at the No.2 tunnel in Nesquehoning, When they worked night shift, they would do their required work that was expected of them and then look for a place to get a little shut eye. One night, he laid down in a room next to where the air compressor was stored. One of his buddies went in and turned the air compressor on, full blast. Lennon was fast asleep dreaming he was on board a submarine when the blast of air went off. According to Lennon, it scared the holy heck out of him.

Getting Even:

This story was told to me by the perpetrator himself. For some reason, two miners had to go up the neighboring chute to get to their chute. They went up the chute and then crossed over by the use of the "monkey" man-way to get their workings. As they were walking through the monkey, the other two miners set off the charge in the face of the chute where they were working. The dust and dirt all blew up into the airway and the monkey. The two miners couldn't see anything and could hardly breath. Chas McMicheal the one miner that was going to his chute told his buddy that we was going to fix their boots. (get even) It seems the night before he was drinking quite heavy and did a lot of eating. At this point; nature called. He had to relieve himself. Most of the time, the miners did this in the water that flowed along the side of the gangway. All the excrements would then flow into a deep sump.

Since the air was pulling up through McMicheal's chute and then down the neighboring chute, he relieved himself right where the air was pulling down into the breast of the miners who set off the charge. When he did this, within minutes the other miners were screaming. They were choking and gagging. They couldn't stand the stench. These guys thought that something crawled up him and died. Chas and his buddy had a good laugh and when they thought that the other guys had enough, he buried the poop with a couple shovels of coal. Whenever Chas. tells this story, he still has to laugh about that incident.

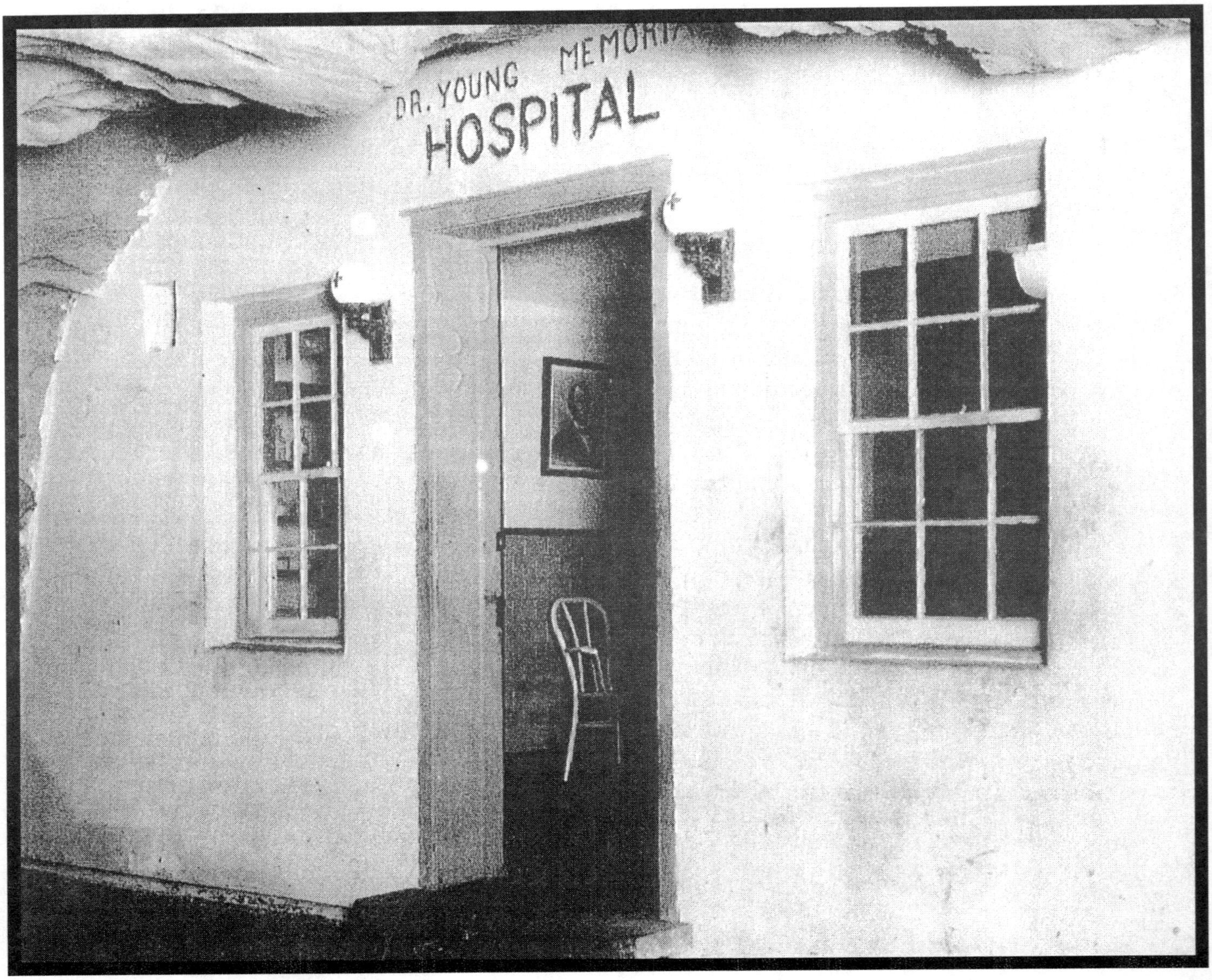

Dr. Young Memorial Hospital room. Built to give first aid to any accident victim before removal from the mine. The room was carved out of a solid rock wall. It is located on the water level gangway of the No.9 Mine.

Breaker boys: Some coal collieries had some of the working areas in the coal chutes made with back and arm rests. Some just had a plank seat where the workers used both their hands and feet to go through the material coming down the chutes. Most were made in such a manner that the "Breaker Boys," stayed bent over all day long.

Mine Mules:

A mule is a cross between an ass, or donkey and a horse which is called a mule or hinny. A mule is the offspring of a male donkey (jack) and a female horse (mare). It is much like the horse in size and body shape but has a shorter, thicker head, long ears and a braying voice of the donkey.

In the early days of mining, a good mule was valued more than a human. Whenever a mule would die or get killed, the Company would fill out a death certificate giving the name of the mule and value. Two such mules certificates at the #9 Mine Wash Shanty are for a mule named Hero and a mule named Spring. On June 12, Hero died. On June 13 Spring died, Value of the two mules was reported at $200 each. In earlier days, it had been alleged that in the event of an explosion, bad mine accidents, underground fire or floods, the Colliery management was more concerned about the fate of the mule more so than that of the worker. It was a known fact that a replacement mule could cost at least $200, while replacing a mine worker was next to nothing. They had a list of people who wanted jobs in the mines and any openings could be filled easily.

Rules On Mule Stables And Buildings Inside The Mines:

It shall not be lawful to provide a stable for a horse or mules inside a mine, unless space for said stable is excavated in solid strata of rock, slate or coal. If excavated in the coal seam, a wall shall be built of brick, stone or concrete not less than eight inches in thickness along the face of the coal from the bottom slate to the roof, or cased in entirely with incombustible material. In the construction of said stable wood or other combustible material shall not be used except for a floor where the animals shall stand upon.

No hay or straw shall be taken into any mine, unless pressed and made into compact bales which shall be kept in a storehouse apart from the stable and built in the same manner as the stable. Under no circumstances shall the hay be stored in the stable. No open lights shall be permitted to be used in any stable, storehouse or at any other place in the mine where the hay and straw are handled. The air current used for the ventilation of a stable shall not be intermixed with the air current used for ventilating any other portion of the mine.

All buildings inside a mine including engine houses pump houses and all other buildings and shanties shall be built of incombustible material which shall be approved in writing by the inspector.

Story About A Mean Mule And A Young Mule Driver Boy:

This story was told by one of the guides at the Scranton Museum Mine Tour. The story goes: A mule driver boy was leading the mule through a somewhat narrow passage, when the mule began to squash the driver against the mine tunnel wall. Being the young driver was pinned and did not have too much room to swing his whip, the mule kept squeezing the young driver. In desperation the young mule driver used the heavy end of the whip handle to hit the mule over its head. The young driver was big and strong and consequently killed the mule with the blow. The mule drivers foreman was called and viewing the dead mule, he said to the young mule driver, "you kin go home now."

"Ahh, an' when do I come back t'work?" asked the boy. "You kin come back t'work when'at mule comes back t'life," said the foreman. At this point, you can see the value of a mule was much higher than the man, in this case a boy. In the Panther Valley Mines, most Mule Drivers were men. Boys were used as slate pickers, door openers, helpers or runners.
The usual procedure for anyone responsible for injury or death of a mule was to be fired on the spot. No if, ands or buts. Caring for mules had high priority in and around any Anthracite mining operations. Most mule drivers had full responsibility for the care and welfare of the
mules in their charge. It was their responsibility to see that the mules were properly fed, watered and rested. Some mining operations had stable bosses that overseen the care of the mules.

My Sweetheart, The Mule: Printed From The Book "Black Rock"

My sweetheart, the mule in the mines,
I drive here without reins or lines,
On the bumper I sit, I chew and I spit,
All over my sweetheart's behind.

This song also appeared in the book, "Black Rock." It was made up and sung by a mule driver by the name of Stutzman, while on the bumper of a mine car, driving mine mules.

When I was just a little boy, I lived in Sarotogy;
I was my mother's pride and joy,
I was always up to roguery,
They tried to educate me, but I wouldn't go to school,
That's why I'm in the coal mine now,
Driving this damn mule.

Mule Was Considered A Very Intelligent Animal:

When the canal systems were first built to haul coal, horses were used to pull the arks or canal boats. They soon found that a horse would pull or work way beyond their means and would literately work themselves to death.

They also found that a mule had a higher value than a horse because they were more sure-footed, more powerful, and way less susceptible to certain illnesses.

Also in the earlier days of the canals, horses had a fondness for water and at times would drag a young horse tender or driver into the canal itself and at times this would cause the death of the young horse tender. They found out mules weren't that enthused about water, and would stay out of the canals. Also an important changeover to mules was that they would never pull any weights or do anything they wouldn't want to do. Thus: they didn't work themselves into the ground like a horse. Between 1827 and 1844, mules were used for hauling the return mine cars on the Mauch Chunk to Summit Hill, Gravity Railroad. When the back track was built they didn't have to pull the cars the entire distance back, so they built special railroad cars for the mules to ride going back to Mauch Chunk. These cars were hooked behind the coal cars. When this back track was put in, they needed 300 less mules.

After a short period of time, the remaining mules became so accustomed to riding that they would never walk back to Mauch Chunk again. The men in charge of the mules would also feed the mules while they rode the Switchback Railroad. This first mule car where the mules were fed, became the first "Railroad Dining Car," in America. How about that!

LC&N Co. Mule Names:

Some of the names of the mules that worked in the Lehigh Navigational Coal Company are the following: Hero, Spring, Eva, Joe, Yankee, Major, Scotty Bessie, Russell, Punchy Harry, Dick, Taps and Spark Plug.

Care and handling of mules was the full responsibility of the driver. The mule and driver had the responsibility of delivering empty coal cars to the vicinity of the miners work area. They also removed the loaded coal cars and transferred them to the foot of the shaft or slope for movement to the surface areas.

John "Scuffy" Yurko of Summit hill was a mule driver when he worked in a section of No. 9 Mine, called, Springdale or "Spring Tunnel." Many of the mule drivers would bring in some sugar, candy, cookies, or apple, treats for the mules. Some mules became fond of chewing tobacco. Many a mule tender bribed his mule with chew. When you gave the mules treats like this, they became your friends. One thing you didn't want to do is get the mules mad at you. You didn't want to mistreat a mule because mules were known to be damn right ornery. If the mule didn't want to move, there wasn't too much you could do to get them moving and a mule never forgot! I must also report that some mules were hard to work with, or as some of the miners called them, "Darn right crazy." Just maybe, they had a reason for their biting, kicking or acting up!

Most mule drivers respected their mules, but every once in a while one was mistreated. The Company certainly didn't condone any mistreatment because each mule at that time was valued at about $200. Which, at that time, was considered a lot of money. You could never overload a mine car with coal. Some mules when they felt the extra weight stopped dead in their tracks. You couldn't make them move. Some mules that pulled three full cars of coal would count the cars as they were hooked up. The mule would hear the clang of the chains being hooked on the mine car. If he heard a fourth clang, he would freeze in his tracks, he wouldn't move no matter what the driver did.

Some drivers got a little clever and hooked the fourth car real quite like, so the mule couldn't hear the clang. The mule would then pull four cars. Most mine collieries had a large mule stable built on the ground surface.

Mine mule being shoed

Mine mules with coal company homes in the background. These homes were built and owned by the coal company and rented to their mine workers. Most of these small settlements were called, "patch towns."

Number 9 had a large barn and yard for the mules. During the late 40's it burned to the ground. Also down in the mines, stable were built for the mules. Some were only two or three stalls, and some were quite elaborate and could have over a dozen stalls. All underground stables were carved right out of the rock tunnels. They were placed for convenience to their work area and so that the mules weren't in any draft or danger.

Mike Sabron from Summit Hill, drove mules for seven years. Two of his favorite mules were Joe and Yankee. He especially liked Joe. In the early days, Sabron would use the mules for the muckers. He would take the loaded cars to the counter chute with the mules. He claims he would dump between eight or 10 cars a day, and then was allowed to go home.

An interesting part was that he was getting a lot of overtime driving mules. A contract miner made about $100 in ten shifts, and he made $187, driving mules. Of course this took in a lot of overtime. Sabron said that his mules would usually pull about 20 empty mine per trip. The mules would pull about 5 trips per shift. Most of the time, they only pulled one full car of coal to the counter chute. One day they gave Sabron a new mule. When he tried to put on the mule's harness the mule pinned him to the side of the rib. The mule started kicking like mad and finally stepped on Sabron's foot. He just kept kicking away and wouldn't stop. Some of the mules didn't like to have their ears touched. When they would try to put on a mule cap, the mules would start kicking and jumping around.

The miners would get a piece of rope, rap it around the muzzle, close to the nose and put in a stick to tighten the rope (like a tourniquet.) They called this a twitch, or cinch. When they did this, the mule would freeze up and just stand there. This gave the miners a chance to put on the mule cap, or on occasions, they used this to shoe a mule. Sometimes this didn't even work! Mike Knies told me this story of when the LC&N Company bought an army mule to use in the mines. When it was time to shoe this mule, nobody could do anything with it.

Down in the Lansford Shops, there were about 14 black smiths that claimed they could do it. None of them had any success. They called Knies's father who was a black smith in Lansford. When he went to shoe the mule, he petted the mule, talked with him a while, patted him down and made friends with him. Before you know it, he was shoeing the mule. Sabron remembers when he got three double shifts, giving him six shifts in one day. It was Sunday work and they wanted to replace a 19'6 timber that was starting to break. They waited an entire shift for the long collar to come and when it came, it was measured wrong. It was 16'9 inches long. This drove the foreman crazy. They stayed until the new timber arrived and put it in place.

Sabron worked a lot of Sundays, fixing oak plank that were used on the counter chutes. Tom Merman of Nesquehoning, who worked in Nesquehoning #2 Shaft, was a mule tender for two mules named Bob and Nick. He said one of the bigger mules could pull three cars of coal at a time. He also said that the mules were very smart and learned repetitious jobs. They knew their routes, they knew what was expected of them. Many mules when bringing a fully loaded mine car to one that is spragged and waiting for a trip to be made up, would step to the side of the rails and let the full car run into the spragged car. In some cases, if the mules didn't step to the side far enough, they would be hit in the hip area. This would usually mean a broken hip and they would have to be destroyed. Merman related a story to me about one of the mules getting electrical burns from coming in contact with the trolley wire. To make a long story short, the electric shorted down through the rigging that was mounted on the mule to pull the mine cars. The two chains that went along the sides of the mule, back to the mine car got so hot that they literally burned the hide of the mule. The burn was so bad that he had to be treated by a Veterinarian and put out to pasture to recuperate for about six months, before returning back to work in the mine. After this, when Merman would see an electric motor car approaching, he would take the mule in one of the shelters or sidetracks and place the mules back end toward the on coming motor. Some times as the motor trolley pole would pass by the hangers there would be a spark. When this mule saw this, it would get very scared and was very difficult to handle. Even with Merman holding on the bridle or bit, the mule would rear up and lift Merman off of his feet.

Merman and other mule drivers have told me that if you saw a mule starring at you, they were out to get you. They would either position themselves to kick you or pin you to the timbers or the side of the mine. Most of the time when this happened, the party getting kicked or pinned probably deserved it. Although at times, some mules were very cantankerous, or had some type of mental problem. Yes, it happened to mules! One mule driver from Coaldale carried a fork along with him. If the mule started pinning him to the sides of the mine he would jab the mule with the fork. This made the mule back off. When Yurko worked at Spring Tunnel, he was in charge of a mule named "Punchy." Yurko claimed that this mule was a good working mule and quite smart. He also said that all the mules were kept down in the mine at that time only brought out when the mines went on their two-week vacation or long strikes.

At the time he worked there, they walked the mules up to the No. 6 mule stable and yard. It did seem that whenever they brought the mules out of the mines, as soon as they got out in the sun they would like to roll around in the dirt and run around and romp. I guess this is the only time they really were free to kick up their hoofs and stretch their muscles. On one occasion the mule took off and they had a hard time rounding him up. At one point they brought up a big red colored mule, named "Scotty," from the 4th or 5th level of the mines. No matter what they did this mule wouldn't do any work. No matter how much they prodded this mule, he resisted. In time this mule got killed when he ran from the mule tender and touched an overhead electric trolley line.

Daniel "Dutch" Yates of Summit hill remembered when "Scotty" worked down on the 4th level. He claimed "Scotty" worked well. Something must have gone wrong when they transferred "Scotty" up to the water level. Most of the mules were treated or dispatched by Veterinarian, Dr. Weaver. Yates remembers Dr. Evans shooting mules that were hurt bad or what ever. Just maybe Dr. Evans came in the mine to treat a human accident victim, who might have gotten hurt along with the mule. Instead of calling Weaver in, he just dispatched the mule himself. The usual procedure for working the mules at Spring Tunnel, was to take the full mine car of coal and dump the coal down a "Counter Chute," which would go downward about 250' to the main tunnel of No. 9 mine.

Second shift crew of No.9 Mine would haul the coal out of the mine Yurko remembers one time when one of the mules touched the overhead trolley line with his ear. The electric current ground out with the mules hoof that had a steel mule shoe mounted on his hoof. The hoof was setting on top of the rail at the time. The mule went down like a load of bricks. The mule tenders thought the mule was dead. About 10 minutes later, the mule got up, shook himself and went back to work, like nothing happened.

Another time, as a motor car was going by, a miner's light shown toward the mule. Mules always go toward any light. The mule ran after the motor car, that was hauling a couple of mine flat cars. The mule jumped up on one of the cars and just stood there. The miners had a hard time backing him off the flat car.

Up at Spring Tunnel, one of the mules fell down the counter chute and hurt itself seriously. One of the other miners carried a 22-caliber rifle to work and would hunt on the way home, from work. They brought the gun in and dispatched the poor mule. When electric motors were introduced into the mine, the mules worked along with the electric motor mine cars. One rule of the road was when the doors of the mine were open, the mule tender, left them open, when he found them closed, he closed them.

Most of the time, the doors were open or closed to help control ventilation into certain parts of a mine. When an open door was left open it could mean that an electric motor with a trip of cars was either coming in or going out of the mine. In 1880, there were approximately 7,000 mules working in or around the hard coal regions of Pennsylvania. One mine mule could handle from 20 to 50 ton of coal during a normal workday. Of course a lot of this had to do with conditions of the mine, rails, and the pitch of the rail bed, and the size and power of the mule.

Mine Workers Killed By Mules:

The following section I have gleaned from the State Mine Inspectors Reports, on how some Miners or Mule Tenders got killed by mules. The job wasn't always as easy as it may appear. Some of the mules were temperamental and had their moments. I guess we can chalk this up to being in the wrong place at the wrong time. Joseph Tirpot, loader was fatally injured by being kicked by a mule. He died January 10, 1905. He had opened the door on the gangway to allow the driver to pass through with a trip of mine cars, and stood on the low side of the gangway with his body close to the edge of the door to keep it open.

When the mule passed him it had its leg over the traces and without any warning kicked with the foot that was over the trace, and struck Tirpot on the forehead. James Shields, a mule team driver that worked in Clarence slope, Clarence Coal Company, was fatally injured on October 12, 1905 by a kick from one of the mules he was driving. He died about three weeks later. John Bonnard, an English, outside mule driver that worked at Lawrence breaker, got on a mule to ride to the barn. About one half mile from the breaker, and when the mule reached the barn Bonnard was found by the barn-boss dragging on the ground dead with his foot fast in the

Miners entering the Spring Tunnel Mine (slope) during the year 1909. Notice the oil lamps on their soft mine caps. This mine was a sister mine to the No.9 Mine. Coal at times was dumped down a counter chute into mine cars in the main gangway of the No. 9 tunnel for delivery to the No. 8 Coal Breaker.

trace chain. He had been clubbing the animal until it became uncontrollable, and in his effort to get it off his foot was caught in the trace chain. Falling from the back of the mule as he went through the streets of the local town killed another mule driver, who worked on the rock banks. Being it was after working hours, this wasn't considered a mining accident.

On May 25, 1912, William Saburis, mule driver, aged 23, was fatally injured as he was coming out of a gangway with a trip of mine cars. The mule kicked him and he fell beneath the mine cars. A John P. Sharpe, Mule Driver, aged 22, was killed at the LC&N No. 8 mines. A mule kicked him. He was driving a three-mule team, in the East Mammoth gangway at the time. His skull was fractured. No one saw the accident happen. He died the next day. On May 8, 1888, John L. Hughes, age 16 and working at No.9, died of injuries from falling off a mule. Quite a few mule drivers were killed throughout the anthracite coal regions by falling under the wheels of mine cars. One bad practice of the mule tenders was to slide one foot along the rail. If the foot got caught in a rail frog, he would be pulled off the car and run over. This type of accident happened every so often. As I read through various Mine Inspection Reports, I saw many injuries that happened to mule drivers. Most non-fatal accidents were from the drivers either being kicked or squeezed by the mules.

A common occurrence was being kicked in the head or stomach. Many drivers had teeth knocked out, fractured skulls, or bad cuts of the face and jaw. Some had fractured legs and jaws from the kick of a mule. Quite a few reports had bad injuries by the mules squeezing the drivers between cars, timbers or the ribs of the tunnels. But overall, I would think that the incidents were minimal when considering them with all the other injuries that happened inside and outside the mines. I mentioned this before, but probably many of these kicks or squeezes were the fault of the mule drivers. Some mule drivers were mean and expected too much from the mule. I would imagine that on some days, the mules just didn't feel too good or had some physical problem, and slowed down, only to be whipped or beat with a sprag or two by four to get them moving. The only defense these mules had was to kick, bite or squeeze the over-bearing mule driver.

Cheating And Kickbacks!

Some of the contract miners who got paid by the lineal foot of coal mined, would use a little chicanery when measuring the coal. This method was also used for miners driving gangway tunnels. If the fire boss was out of sight or with his back turned, the miner would take the end of the tape to the face of the coal vein. They would fold over the tape a few times so they got extra yardage. Another thing that was done to fool the bosses was when scooping up coal from the gangway tunnel floor, the workers who got paid by the car, would dump some old mine timbers or planks on the bottom of the mine car. Then shovel the coal on top of the junk. This way they got their quota of cars filled faster and had a chance to go home earlier.

Many fire bosses knew that the miners were cheating. Some just overlooked it, while some fire bosses wanted a piece of the action. Many of these bosses received pay off, or kickback as it was called. The miners would hang their large overcoats on a nail, going up into the airways. In the inside pocket, they would place a pack of chewing tobacco. Inside the bag they would put kickback money for the fire bosses. Whenever a fire boss would go up into the chute to check on the miner's progress, he would go into the pocket and take his ill-gained booty. When the miners saw him, and if he was smiling, they then knew he was happy with his pay off. One particular loader saw what was going on, so before the fire boss made his rounds, he went to each coat and retrieved the money. When the fire boss checked the coats there were no pay offs.

When the fire boss went up the chute, the miners could see something was wrong. They asked what was the matter? At that point, they asked if he wasn't happy with what he received? It didn't take long before he smelled a rat. He knew someone beat him to the money. He did tell everyone that if he finds out who did it, he would have him fired. In the same case, he himself placed his own job in jeopardy by lying about the lineal feet of coal or gangway driven and then taking kick backs. Many of the miners would meet the fire boss in the local gin mill on payday. The miners would treat the boss to several rounds of drinks. If the miner made extra money on cheating, the fire boss would get himself at least $15 from each contract miner. This way he made himself a nice haul, from the miners who went along with this sort of thing. Many of the miners were hunting buddies or friends of the family and the fire boss would falsify the records just to help out his mining buddy.

One case in point was my own father. His fire boss and his wife were good friends of our family. We would go down to Seek and visit them on holidays and other occasions. His wife was a very sweet lady and as a child I enjoyed her company. Her husband the fire boss was another story. One year, the week before Christmas, my father came home with a pay check of 50 cents. The company wasn't working full time and he had coal and other expenses taken out of his paycheck. Being short of cash, he asked my mother for some extra money that she had put aside, to buy his fire boss, and friend, a box of cigars and a bottle of whiskey for the holidays. When my parents went down to their home, he opened up the box of cigars and looked at my father and asked him if he didn't forget something. My mother was in ear shot and came back at him that all we had left in the pay check was 50 cents, and what he saw was what he was getting, and that he should be glad that he got that.

Knowing my mother had heard his comment, he turned around and laughed, and made a big joke of the matter; like he was just kidding. My mother knew better! In the same case, many miners wanted no part of cheating and did everything on the up and up. Some miners were known to cheat the other contract miners by switching tags or chalked on numbers on the mine cars that were filled with coal from someone else's chute. The Coal cars would be hooked into a trip to take out of the mine, when another miner would take off the tag, or rub off the chalked on number on the car. They would replace that number, with their number. This way they got paid extra money for the extra cars that they supposedly produced. Some fire bosses would also take any extra cars that the loaders filled beyond the day's quota of cars and park them overnight in an unused gangway. This way, if he didn't get his quota on another day, he would use these cars to make it up.

In the same case, they also juggled the tags so that if one of the miners didn't have a good day, his quota would be made up with these cars. This worked at times, to everyone's satisfaction.

One of the biggest cons to the mining company happened up at No. 6 mine and breaker in Lansford, Pennsylvania. This happened by accident, but after a while I am sure that the miners knew what had happened. They just took advantage of a good thing.

As their chute and breast reached up close to the surface of the ground, the miners broke through into a large culm bank. This is the bank where coal and slate refuse gets deposited. All this refuse was from years past. In all reality, the company paid miners for the original mining of this material. For months, the miners poured this material down their chutes, into mine cars, and then to the breaker for preparation. After a while, someone took notice that this material coming down the chutes just wasn't natural and something was a miss. Once the engineers started checking it out, it was quite obvious what was happening. In fact, anyone traversing the Summit Hill to Lansford highways could see a large indentation in the culm bank. Each week the indentation kept getting bigger.

What happened, is the Coal Company paid twice for this material, and certainly didn't salvage too much quality coal from the old culm (refuse) banks.

One thing we do learn in life and that is if there is some way to beat the system, somebody will come up with it.

Nick Perbetsky, from Summit Hill, told me he would bring a jug of wine to work every so often. Some days he would take it in the mine and hide it behind a timber. Every so often he would take a slug. He said when they got their quota of cars out, they would end up playing pinnacle and drinking wine. One day he was tantalizing the hell out of one of the guys when this guy pulls a knife on him. All of the other (continued page 44)

Two mine mules exiting a mine slope. Miner has a soft hat with an oil lamp, Oil lamps caused many fires and explosions before switching over to a battery lamp, which is still used today.

Young Mule Driver: Most mule drivers were at least 14 year of age. Some were younger.

Case No. 234- Edward Cannon Irish laborer, aged seven years Was fatally injured by mine Cars in the No. 5 mine in Lansford, PA on the 26th day Of November, 1884.

The young man was engaged in driving mine cars (driving mules) this day and while taking out a trip of loaded mine cars, slipped and fell under the cars and had his leg and hand badly crushed, and from the effect of which he died.

miners grabbed a hold of the knife wielder before he caused any damage. I guess that sometimes we all tend to go too far!

In the earliest of years, men pushed the mine cars in and out of the mines, in time oxen and then mules were introduced into the mines. The mules that were in the Panther Valley mines, lasted until the early fifties. At that time the Company sold off all the mules. A pair of mules sold for $500.

Mine mules in a pasture at one of the coal company farms

Decline Of The Mining Industry:

From 1902 until World War I, the hard coal industry enlarged and had some of their highest production. During the 20s, the industry had more than its shares of work stoppages or strikes. When you think of it, each strike from that time on was a nail in the proverbial coffin.

The Coal Industry raised its prices constantly, the Railroads had high shipping rates for the transportation of coal to the city markets. Between lower production, and the shortage of coal during longer strikes, people or industry started looking for alternative fuels. Oil, gas, coke, electric and in time, nuclear fuel, started coming into the picture. As people or industries converted to these new fuels, there was no turning back. Between 1934 and 1936 "Equalization," between the coal companies was being pursued. This would give everyone equal working time. This was great for the working man, but not necessarily good for the coal companies.

Certain operations that weren't profitable, would remain working, while operations producing the most coal would be limited to operations. Steel Mills started using more coke instead of Anthracite Coal. During the 20's, several long strikes took place. The miners asked for certain things, while the Coal Companies wanted wage reductions. Neither side won, and when the strikes were over, each side claimed victory. In all reality it was a meaningless victory for all.

The miners were pushed for more production. Conditions worsened. Having a good paying breast or chamber took payoffs to local foreman. The union officials backed off. Even during the 30s, my own father, had to promise to buy a car from one of the mine supervisors who owned a garage, so he would give a job to one of his relation. This went on all the time. By the later 30's, coal production was reduced to all time lows. Work around the collieries, went down hill, until the mines finally closed. In 1954, the LC&N Co. closed down all its deep mines. After six months of hiatus, mining resumed as two local firms, The Panther Valley Mining Co. and the Coaldale Mining Co. In 1960 these operations were closed. Fauzio Brothers leased (1960) and then bought the companies mining lands in 1966 and built a new modern breaker. In time, many of the large coal breakers where eventually torn down. These were the last monuments of when "Coal was King."

It was the unions that brought the worker out of the disparages of those earlier years. Between greed from the Coal Barons and all the absentee land lords...the high union demands for higher salaries and benefits, "King Coal," got enthroned. It was everybody's fault and yet no ones fault. I guess we can call this, "The price of progress." New cheaper fuels, plus the fact that they were more convenient, broke the back of the coal industry. Most of the Anthracite Coal produced now is mostly strip mined. With large shovels, giant sized trucks for hauling the coal, modern breakers and fewer workers. Production and sales seemed to have leveled off, but is still considered a viable industry. There is a story to be learned. First, our natural resources are limited and precious. With new discoveries or inventions, the world can and will change. Where oil or gas took over for coal, Hydrogen or some other similar product will take its place as a cheaper and more efficient fuel in the future.

Epilogue:

In June of 1972, all deep coal mining came to an end in the Panther Creek Valley. It was the coal industry that fueled the industrial revolution in our country. But as some things are born, some shall also perish. Before the exploitation of all the coal fields in Pennsylvania, the forests were lush and green. and Virgin stands of timber were in abundance. Once the coal operations started, nature took one a terrible toll. The land was laid waste. Both land and water were desecrated. The human toll from mining was intolerable. In all of the anthracite coal fields, over 31,000 mineworkers met their death over an 108-year period. There were also many deaths in the earlier years that weren't recorded.

This was a high price to pay for the exploitation of coal. We also should not forget the large number of injuries and incapacitation's. Researchers claim that close to 45 injuries happened to every fatality. Along with these dreadful statistics is that every year since the closing of the mines, many miners die or are suffering from the dreadful black lung disease, Anthrasilicosis. The unofficial death toll of the Lehigh Coal and Navigational Company from the 1900s to the closure of the mines was 456 deaths. Death at times came too easy. No one can even begin to write about the untold sufferings of not only the miners but of their families. Life itself, around the coal regions, was never easy.

According to the records; Number 9 mine had 13 deaths, since the early 1900s. This doesn't take in any deaths between 1855 and 1900. Probably the reason for the smaller amount of fatalities is the fact that the main tunnels are carved through solid rock. Andrew Boretsy was the last man killed in the No. 9 mine, in 1959. He died from a rock fall.

The quote below comes from George Korson's book, "Black Rock"

Forty years I worked with pick and Drill
Down in the mines against my will,
The Coal Kings slave, but now it's passed;
Thanks be to God, I am free at last.

The above poem came from a tombstone of an anthracite miner in St. Gabriel's Cemetery, in Hazleton, Pennsylvania.

And God shall wipe away all tears from their eyes; and there shall be no more death, neither sorrow, nor crying, neither shall there be any more pain; For the former things are passed away. Rev. 21:l

Beginning Of A New Era:

We must never forget that, "Coal was King," It is a legacy of the hard coal regions, which should never be forgotten. People who live or had lived in the coal regions, should be proud of this heritage, and should never be offended when referred to, as a "CoalCracker."

In 1972, deep coal mining in the Panther Valley came to an end, but now we welcome the rebirth of the No. 9 mine, but not for sorrow or death, but for the education of the young, a reminder of the past, and for the reverence of those who paid the supreme sacrifice, down through the years.

A Coal Miner's Reward:

Coal miners are a rugged individual, who lived hard, worked hard and at times drank hard. I think that I could say with complete honesty that when the Coal Miners depart this earth, and when they meet St. Peter at the Golden Gates, he would tell them, that there is only one place for them to go and that is Heaven... For the Miner has surely served his time in hell.

Founding Of The Panther Creek Foundation...Commonly Known As The Wash Shanty Mine and Museum:

During 1990, a group of dedicated individuals met in Lansford, Pennsylvania to start a library at the old No. 9 Wash Shanty building. In 1992, the group obtained a lease from Mr. James Curran who was president of the Lehigh Coal & Navigational Company for the Wash Shanty building and several acres of land that comprised the old No10 Mine Complex. The start up was slow and tedious. The first job was to clean out the old building, which was used as a warehouse. Then some changes to the structure had to be made. Little by little, various people and organizations donated mine artifacts of every description from mine hats to lamps, tools and safety lamps. The mine now boasts of a large collection of one of a kind mine artifacts and some date back into the 1800's. Along with the artifacts, there is a large assortment of informative books, maps, mine equipment and models of all the early mine breakers through the Panther Valley.

In 1994 Attorney James Curran, President of the Lehigh Coal &Navigation Company, gave the Number 9 officers and volunteers' permission to re-open the sealed Number 9 Mine in Lansford, Pennsylvania.

Rebirth Of The Number 9 Mine:

Because of a huge concrete seal that was placed in the opening of the mine portal, a turbo jackhammer had to be secured to breach this wall. When the concrete seal was breached, a large volume of water that was trapped behind this wall flowed out and flooded all the area in front of the mine opening. Because of this, tons of mud was deposited throughout the tunnel. The mud averaged from two to three feet in depths at certain areas. All the cleanup work was done with (6) volunteers on mostly weekends. The cleanup was overwhelming at times. Sheer determination got the main tunnels open to drive a mine motor and mine cars into the mine to help with the cleanup.

In March of 1997, work was then concentrated on opening a second exit or escape way. This was done because of State Mine Regulations. This escape was driven in a five-foot wide vein of coal. Along with the coal, another foot or two of rock was taken out. The first section of escape way is driven at a 45-degree pitch. The second is about a 56-degree pitch. The overall length is about 240 feet long.

After about 18 months of weekend work, the escape way was completed on Thanksgiving weekend of 1998. In the winter and spring of 1999, the old No.9 Mine shaft which was jammed with old timbers that were dropped down the shaft when the coal company placed a seal at the ground level of the old head frame, was cleaned. Because of the huge size of some of the timbers, it was a very slow process. Chain saws and overhead chain blocks had to be used at times to pull and remove these timbers. Safety was a very big issue when it came to removal of this logjam.

Now, the original mine cages (elevators) are fully exposed. These were locked into position on the water level when they filled the lower section of the shaft with shaft timbers. The shaft and cages will be one of the high light of the mine tour. On either side are the old go-devils that pulled cars onto the mine cages. No other mine tour has a shaft or mine cages to show the public.

Along with this work, the clean up of old mine timbers and the re-timbering of a couple of exposed veins of coal where the main gangway tunnels were driven through. The final cleanup of mud and debris was done in the summer and fall of 1999. Where the mud was removed, stone was put into place between the two rails and the rib. This section will be for a walking tour by those visiting the mine. At first, the volunteers worked with very limited funds or equipment. The on-going joke was that the volunteers worked just like the Flintstones, but with some donations and small grants, progress was made over the years.

Now, with the help of state and county grants along with large private donations, the mine crew is able to obtain better equipment and tools to handle the job. With luck and determination, the No.9 Mine will be opened on a limited basis for tourism in the year 2000. This mine has several features that other mine tours don't offer. There is the original hospital room, hewn out of solid rock. In the branch tunnel there is the No. 9 mineshaft that goes down four and a half levels. In the shaft are the two original cages which were the elevators for taking men and mine cars to and from the lower levels to the main gangway. On either side of the shaft are the original go-devils, which are comprised of motors, pulleys and cables.

Last but not least the public will see the original mule way in which mules were brought up from the second level to the first level. Because of these added features the NO. 9 Mine tour will no doubt become one of the State of Pennsylvania's top tourist attractions.

The last pages in this book give you, the reader a pictorial review of what the mine volunteers encountered and did to bring back to life, this "Grand Old Matriarch," of the Panther Valley, so that future generations will have a little understanding of what deep mining was all about.

List of Inside Rates, LC&N Co. (Prior to May 28, 1913)

Occupation	Rate
Battery Starter	.256
Bratticeman	.256
Bratticeman Helper	.203
Carpenter	.256
Carpenter Helper (On Shaft)	.22
Carpenter Helper	.203
Door Boy (In Shaft)	.109
Door Boy (In Plane Level)	.104
Driver 3 Mules & Loading (Water L.	.22
Driver, 1 Mule	.107
Driver, 3 Mules	.203
Driver, 2 Mules	.185
Driver (At Top Of Plane)	.203
Electrician	.256
Electrician Helper	.203
Engineer-Fan, Locomotive, Water L.	.152
Fan Boy, (In Plane Level)	.104
Fan Boy (Shaft Level)	.109
Fireman Locomotive	.176
Laborer, (Contractor)	.213
Coal Pusher (Handling Buggies)	.213
Coal Pusher	.203
Laborers, Bonding Rails	.203
Laborer, Cleaning Roads	.203
Laborers Rep. Airways, Chute, Gang	.203
.167	
Laborers- Company Robbing Gang	.203

Occupation	Rate
Lamp-man	.152
Loader	.22
Loader Boss	.256
Loader Boss & Robbing Boss	.30
Miner Consideration Work	.30
Miner Driving Chute	.256
Miner Driving Gangway	.256
Miner Rep. Airway, Chutes etc.	.256
Miner Robbing Pillars	.256
Motorman-Repairing	.165
Motorman-Running	.249
Plane Tender Car Runner T.P.	.22
Plane Tender Car Runner Bot.	.22
Plane Tender (Top Man)	.256
Plane Tender (Bottom Man)	.22
Plane Tender Empty Car Hoist	.203
Poleman Repairing	.145
Poleman Running	.172
Shaft Tender- Bottom Man	.22
Shaft Tender Top Man	.203
Shaft Tender- Bottom Man	.22
Shaft Tender, Top Man	.203
Shaft Tender Car Runner	.22
Stable Boss & Blacksmith	.256
Stable Boss & Blacksmith Hlpr.	
Track-man	.256
Track-man Helper	.203

Small pond of water that formed in front of the three-foot thick concrete seal that was placed in front of the No.9 mine portal to keep out trespassers. When the seal was breached a tremendous amount of water flowed out of the opening and flooded the low areas outside the mine

No.9 Mine volunteers replacing overhead timbers in a double gangway north of the No.9 mine shaft.

Concrete seal and dirt mound removed from the portal of the No.9 Mine in the year 1993.

Several volunteers removing large concrete chunks from the three foot seal that was placed in front of the mine opening.

Five of the mine volunteers that worked in the No.9 Mine to refurbish and bring it up to state specs for tourism. From left to right: John Fauzio, mine superintendent, Greg Yurchak, Mine Foreman, Dave Kuchta, Jack Julo and Dan Farber. Dave Zabian was absent at the time the picture was taken.

Kuchta and Julo are holding the first large piece of coal that was brought out of the No. 9 mine since it was last worked in 1972. This coal was a part of the second exit or escape way that was needed to comply to State and Federal laws.

Mine motor in the gangway tunnel of the No. 9 Mine. (Cleaning mud)

Two miners exiting the "Mule Way." This mule way was used to bring the mules up from the second level to the water level of the No.9 Mine, in the early years of operation.

Greg Yurchak, mine foreman, dressing off the face of a coal vein that was used for the second exit or escape way.

Greg Yurchak and Dan Farber at the breast of the coal vein that was utilized for the second exit.

Reconstruction of the portal of the Number 9 Mine entrance in Lansford, PA. Dave Kuchta (author) is on the scaffold, Jack Julo is on the ground on the right side. Dan Farber, mine volunteer on the left side of the picture. Absent from the picture are Andy Solomon and Bernie Babinetz who also worked on the portal reconstruction work.

Black Diamonds:

Coal flowing down a mine chute into a waiting mine car in the No.9 Mine

Man way of the second exit or escape way located in the Holmes Coal Vein. Lower section (100ft.) is driven at a 45 degree pitch.

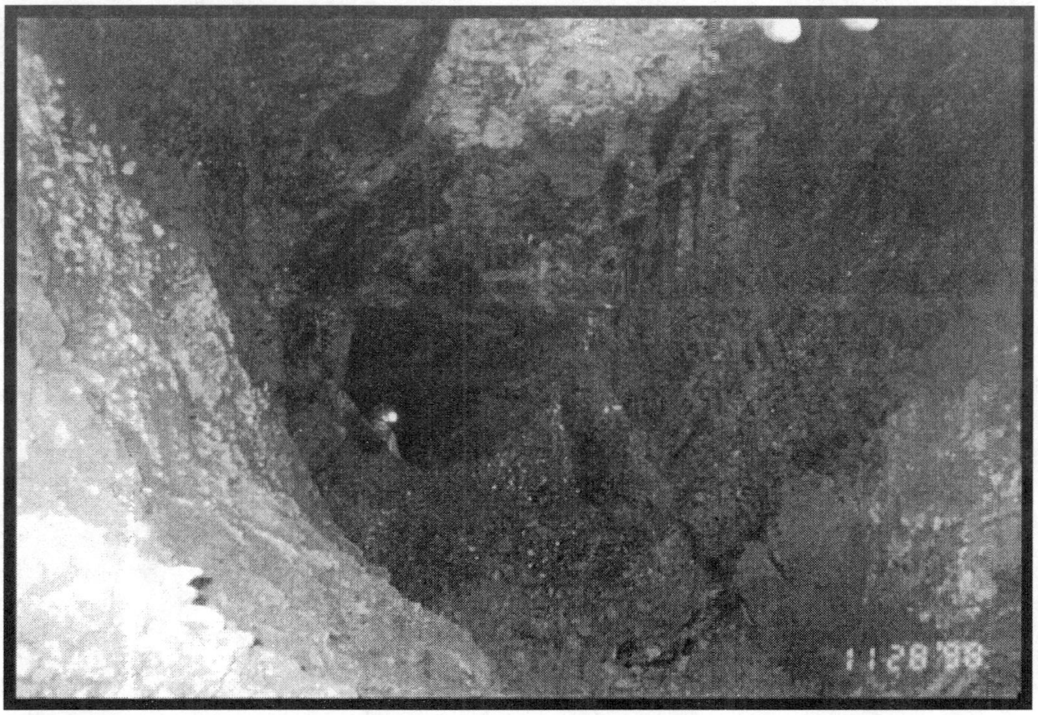

Breaking through to the surface with the second exit or escape way.